The Girls Who Could Fly in the Night

An Inspirational Tale about the
Women of World War Two

Also by Maya Cointreau

The Girl Who Could Heal Your Heart
The Girl Who Could Sing with the Birds
The Girl Who Could Dance in Outer Space
The Girl Who Could Talk to Computers
Gesturing to God: Mudras for Well-Being
The Comprehensive Vibrational Healing Guide
Shamans Who Work with the Light
The Healing Properties of Flowers
The Practical Reiki Symbol Primer
Simple & Natural Herbal Living
Conversations with Stones
Natural Animal Healing
Equine Herbs & Healing
Grounding & Clearing

The Girls Who Could Fly in the Night

Maya Cointreau

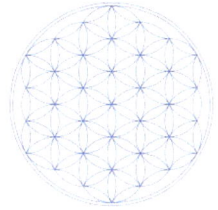

An Earth Lodge® Publication
www.earthlodgebooks.com
Roxbury, Connecticut

All rights reserved, including the right to reproduce this work in any form whatsoever, without written permission, except in the case of brief quotation in critical articles or reviews. For more information contact Earth Lodge®, 12 Church Street, Roxbury, CT 06783 or visit Earth Lodge® online at www.earthlodgebooks.com

Copyright 2017 by Maya Cointreau
Printed & Published in the United States by Earth Lodge®

ISBN 978-1-944396-51-0

All Artwork, Layout & Design by Maya Cointreau

"Know yourself and you will win all battles."

Sun Tzu

A war was afoot.

There were armies of bullies.

The world needed heroes,

Good hearts unsullied.

Most soldiers were men;

It's the way things were.

Women nursed, worked factories,

But they didn't fight wars.

It was 1941

When everything changed.

On one man's orders,

The 588th was arranged.

Girls would become heroines,

Receive honors and ribbons,

In a night bomber regiment

Run completely by women.

From the entire Soviet Union,

Young girls in their teens

Volunteered to run missions

Like night swallows, unseen.

Mechanics and pilots,

Commanders and crews:

All of them girls

Who could do their job true.

Their planes were old,

Made of canvas and wood,

But the girls hit their targets

As only they could.

For those old planes were quiet.

They moved like the wind.

No pilots could catch them

When they idled and hid.

Each plane flew low

Eight times a night.

Two bombs aboard,

Two girls per flight.

Marina, Natalya,

Tamara and Vera.

Remember their names,

For each was a hero.

Fearsome as witches

Flying the night,

They protected their people

From the meanness of might.

They stood up to bullies,

The people most feared,

Lighting hope in the dark

Each time they appeared.

Can YOU be a night witch?

Can you do what's right?

When you stand up for good,

You'll always shine bright.

More About The Girls Who Could Fly in the Night

In 1941 the Soviet Union was embittered in a constant battle against the invading Nazi forces of Germany. More troops were needed to push back the Germans, so Joseph Stalin issued an unusual order. He called for three air force units to be deployed – each of them comprised of women. The 588th Night Bomber Regiment of the Soviet Air Forces, later renamed the 46th "Taman" Guards Night Bomber Aviation Regiment, was born. From its mechanics to its highest commander, the regiment remained entirely female.

All under the age of 26, most only 17 to 22 years old, the women quickly proved themselves as skilled flyers and tacticians. They flew over 30,000 missions in four years, dropping over 23,000 tons of bombs from Polikarpov Po-2 biplanes made of wood-and-canvas. The old planes were designed for crop dusting and training purposes, but the Soviets put them into action. Each plane could carry just two bombs, one per wing, with no room for radios or parachutes. The obsolete planes seemed ill-matched for German fighters. Indeed, the enemies' aircraft would stall trying to fly at such low altitudes and slow speeds – a fact that proved the regiment's saving grace, allowing them to easily outmaneuver their foes.

A Polikarpov Po-2 , Photo by Douzeff, CC BY-SA 3.0

Even so, flying at night into enemy territory, being chased by screaming metal Messerschmitt Bf 109 and Focke-Wulf Fw 190 planes – war was serious business. The women had no rader, just maps and compasses to navigate the freezing darkness as they flew in open cockpits. The planes would be sent on anywhere from eight to eighteen missions each night. There was no rest for the women of the 588th. On average, each pilot flew 800 missions. Thirty of them died in combat. Still, the girls stayed strong and embraced their feminine power, obeying their top commandment: "Be proud you are a woman." They decorated their planes with flowers, did needlepoint in their spare time, and used their navigation pencils as makeup.

The aviators of the 588th quickly became well-known to both Soviets and Germans and finished the war as the Soviet Air Force's most highly decorated regiment. Among their fellow countrymen, the girls were dubbed the "Night Swallows." The Germans did not feel as kindly towards the women taking out targets night after night. Each time the women neared, they would cut their engines and glide below enemy radar to release their bombs. The quiet sound of their planes reminded the Germans of witches on brooms so they called them the Nachthexen, or "Night Witches." Undetectable except by searchlight or the sound of the wind on their wings, the Po-2s almost always hit their marks. Then they would turn their engines back on and escape, out-flying any who dared to follow. Their attacks were so feared that the Germans put a high price on their capture – if you could take down a Night Witch, you'd automatically receive an Iron Cross, Germany's prestigious military honor.

In the end, the Night Witch regiment was disbanded six month after the end of the war. The women were sent home, heroes all, to return to their normal lives. But the Night Witches live on still in our memories, shining examples of true bravery and honor.

Historical sources: Wikipedia, SeizeTheSky.com, History.com, *Curiosity* and *The Atlantic*.

About the Author

Maya Cointreau has been writing and drawing all her life. She lives on a farm in Connecticut with her family, a pack of poodles, a herd of horses, a flock of chickens, and a cartload of cats.

For more information about her other books and CDs, visit her website at http://www.mayacointreau.com.

www.ingramcontent.com/pod-product-compliance
Lightning Source LLC
LaVergne TN
LVHW072127070426
835512LV00002B/31